Overview: Hurricane Damage is Estimated Under Two Scenarios

Scenario with climate change only

- Rising sea levels, which lead to more damage from storm surges

- Changes in expected annual frequency of hurricanes
    - Hurricanes are classified in Categories 1 through 5, with 5 being the most intense
    - Many models predict increases in Category 4 and 5 storms in the North Atlantic Basin (though there is much uncertainty)

Scenario with climate change and coastal development

- Climate change
- Increases in property exposure

Overview: Method and Reported Outcomes

- A Monte Carlo simulation is used to estimate future hurricane damage.

- 5,000 simulations are used to capture uncertainty in factors that affect hurricane damage.

- Annual results include:
  - Distribution of damage estimates
  - Expected damage (mean of estimates)
  - "Likely range," indicating the range around the mean that contains two-thirds of the estimates

Preliminary Results

- Compared with current conditions, expected hurricane damage in 2075, measured in 2015 dollars, would:
  - Double under the scenario with climate change only
  - Increase five-fold under the scenario with climate change and coastal development

- Hurricane damage is projected to grow more quickly than GDP under scenario with climate change and coastal development; in 2075:
  - Expected damage as a share of GDP would be roughly 40 percent higher than under current conditions
  - But dollar amounts would still be small relative to GDP; increase in expected damage would be less than 0.1 percent of GDP

- Estimates are uncertain and likely range grows substantially over time.

Damage Function Used in this Analysis

- Damage function from Risk Management Solutions (RMS) provides state-specific damage estimates.

- Each estimate is a function of:
  - Frequency of landfall anywhere in the U.S for each category of hurricane
    (f_c, where f = frequency and c = hurricane category, 1 through 5)
  - Probability of landfall at various locations for each category of hurricane, conditional on any U.S. landfall (Estimated by RMS on the basis of more than 100,000 hurricane season simulations)
  - Sea level in each of the 22 states included in this analysis
    (s_i, where s = sea level and i = state, 1 through 22)
  - Valuations of current property exposure for each of the states

Preliminary Damage Estimate in Reference Case

- The reference case is estimated damage under current conditions (with no additional climate change or coastal development); it is based on estimates of current:
  - Hurricane frequencies (average over the past 100 years)
  - State-specific sea levels
  - Valuation of property exposure by state

- Reference case estimated damage is $29 billion (2015 dollars)

- Estimate reflects average conditions; actual damage could be more or less depending on actual hurricane occurrences and locations of landfall.

Approach for Estimating Effects of Climate Change Only in Selected Years (e.g., 2025)

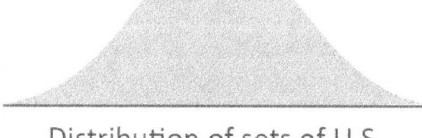

Distribution of sets of U.S. hurricane frequencies in 2025

Random draw

Begin simulation

Random draw

22 state-specific distributions of potential rise in sea level in 2025

Damage Function

Calculates expected property loss in each state given...

- State's property exposure in the reference case

- Frequency of hurricanes making U.S. landfall

- State-specific sea levels

A single estimate of expected loss in each state in 2025

Creation of single point on distribution

Distribution of expected losses due to hurricane damage in the U.S. in 2025

Repeat 5,000 times

Distribution of Projected Sea Level Rise in Two States: Florida and Texas

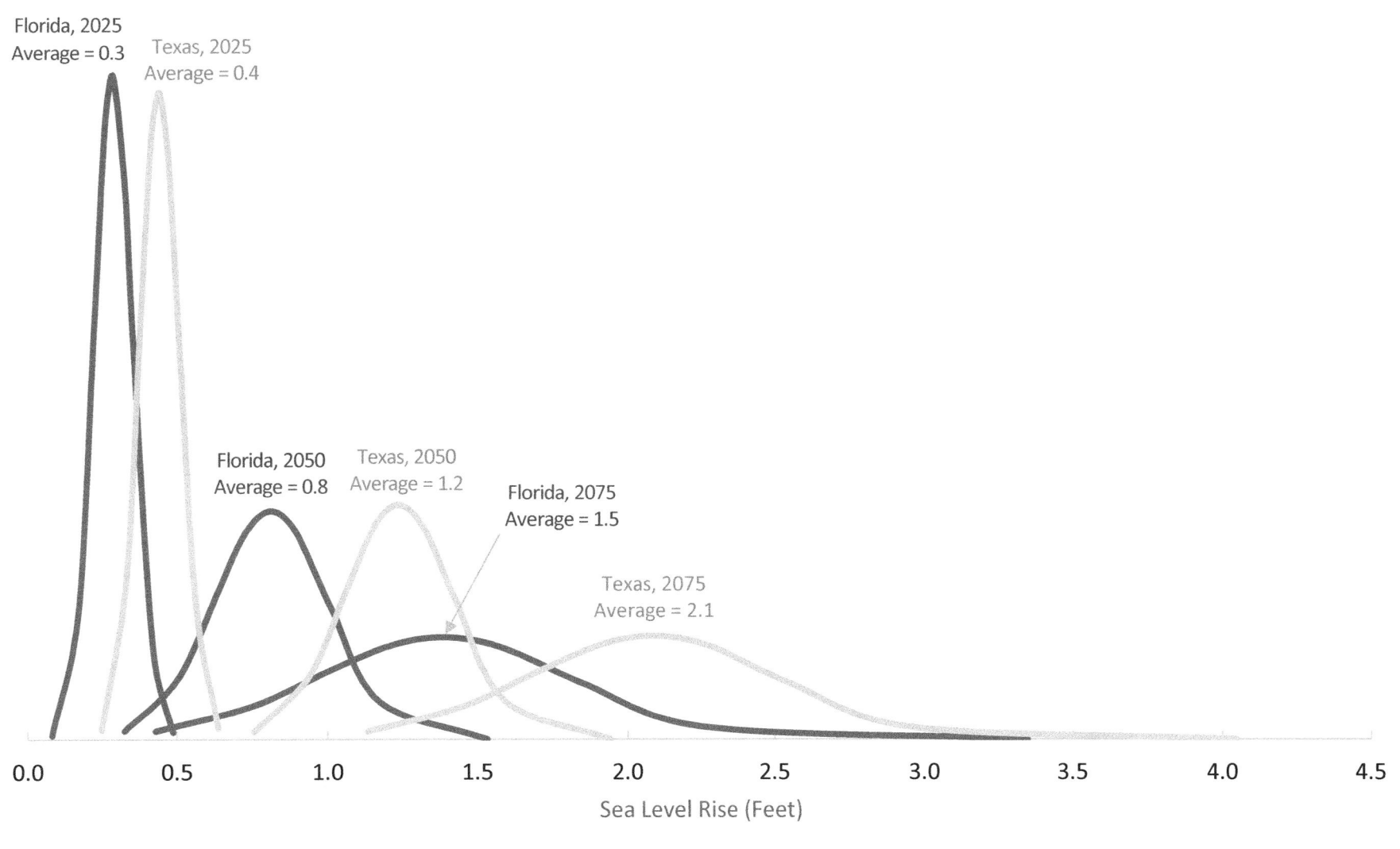

Florida, 2025
Average = 0.3

Texas, 2025
Average = 0.4

Florida, 2050
Average = 0.8

Texas, 2050
Average = 1.2

Florida, 2075
Average = 1.5

Texas, 2075
Average = 2.1

Sea Level Rise (Feet)

Projected Frequencies of Landfalls of Category 2 Hurricanes, Estimated by Two Modelers

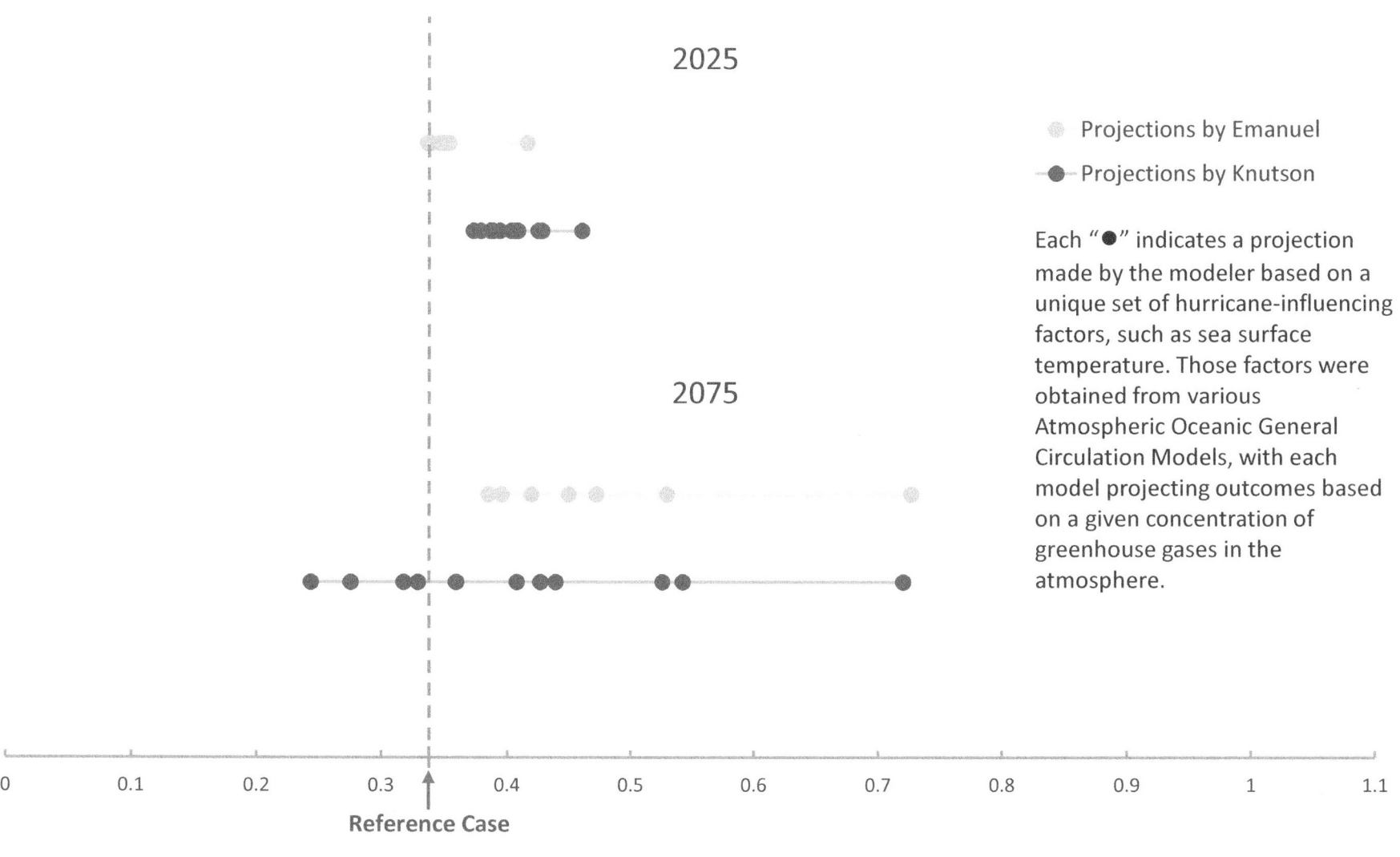

2025

2075

Projections by Emanuel

Projections by Knutson

Each "●" indicates a projection made by the modeler based on a unique set of hurricane-influencing factors, such as sea surface temperature. Those factors were obtained from various Atmospheric Oceanic General Circulation Models, with each model projecting outcomes based on a given concentration of greenhouse gases in the atmosphere.

| 0 | 0.1 | 0.2 | 0.3 | 0.4 | 0.5 | 0.6 | 0.7 | 0.8 | 0.9 | 1 | 1.1 |

Reference Case

Projected Frequencies of Landfalls of Category 4 Hurricanes, Estimated by Two Modelers

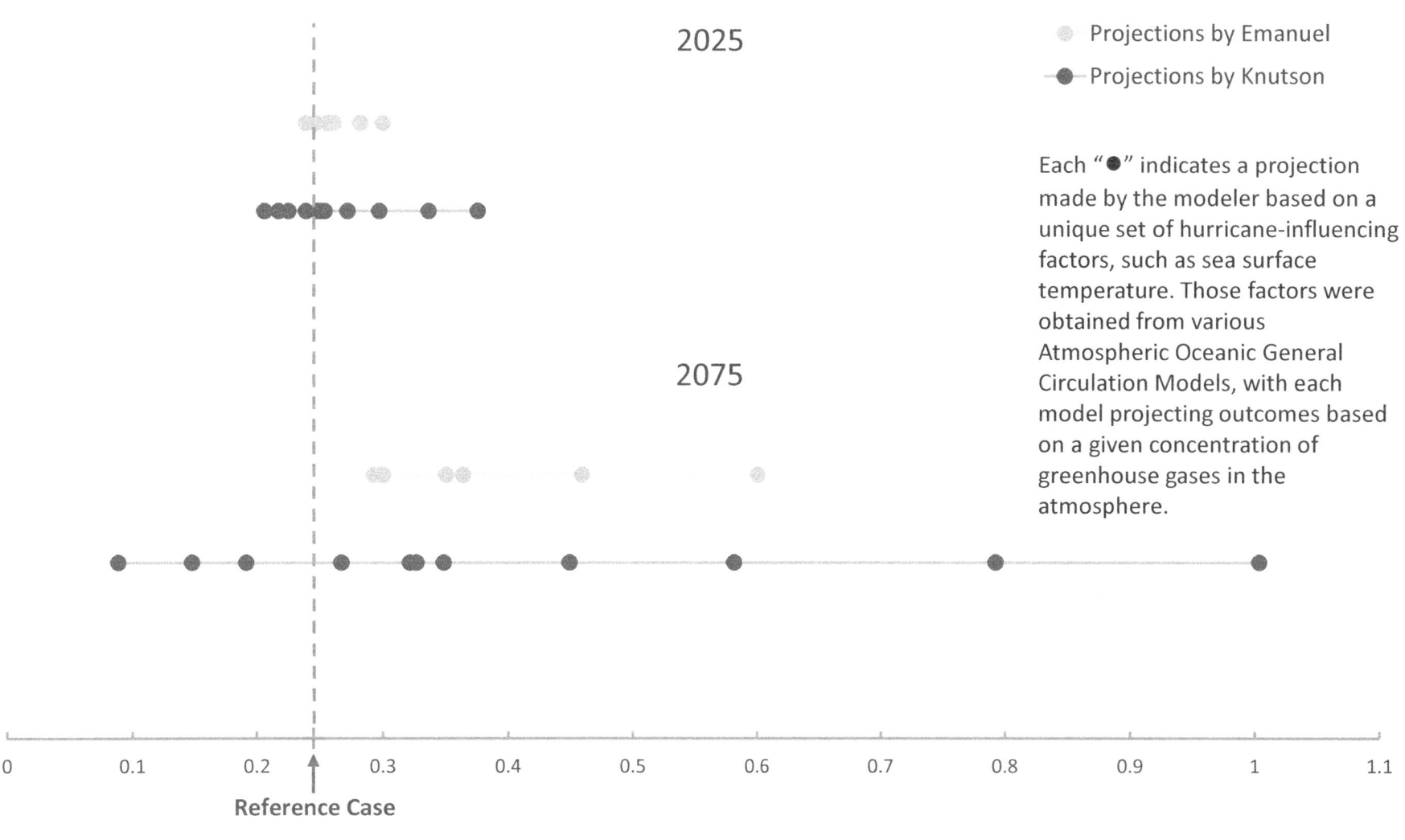

2025

Projections by Emanuel

Projections by Knutson

Each "●" indicates a projection made by the modeler based on a unique set of hurricane-influencing factors, such as sea surface temperature. Those factors were obtained from various Atmospheric Oceanic General Circulation Models, with each model projecting outcomes based on a given concentration of greenhouse gases in the atmosphere.

2075

| 0 | 0.1 | 0.2 | 0.3 | 0.4 | 0.5 | 0.6 | 0.7 | 0.8 | 0.9 | 1 | 1.1 |

Reference Case

Preliminary Damage Estimates, by Dollar Amount, Under the Scenario With Climate Change Only

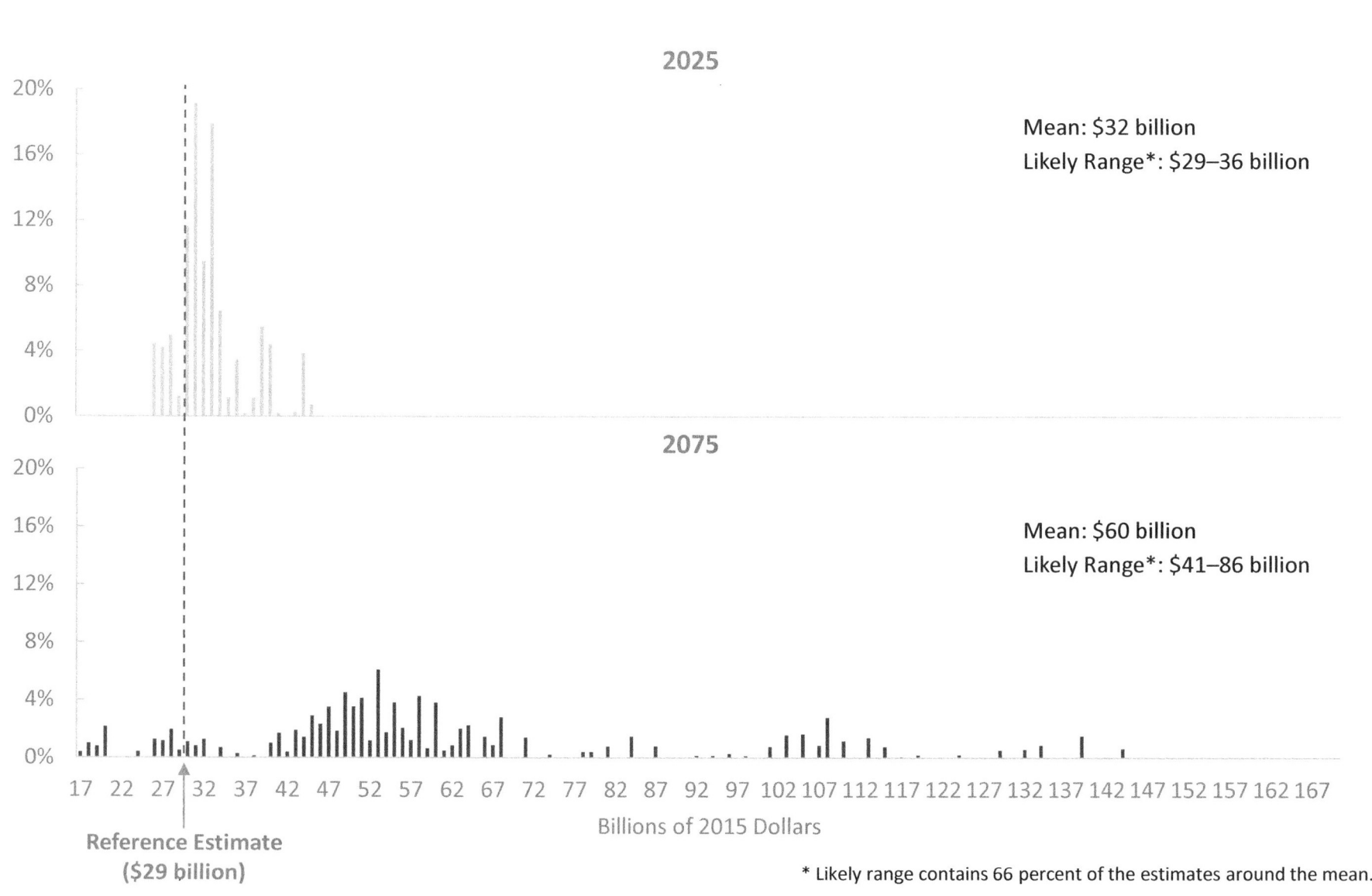

2025

Mean: $32 billion
Likely Range*: $29–36 billion

2075

Mean: $60 billion
Likely Range*: $41–86 billion

Billions of 2015 Dollars

Reference Estimate
($29 billion)

* Likely range contains 66 percent of the estimates around the mean.

Estimating Effects of Climate Change and Coastal Development in 2025, Florida Example

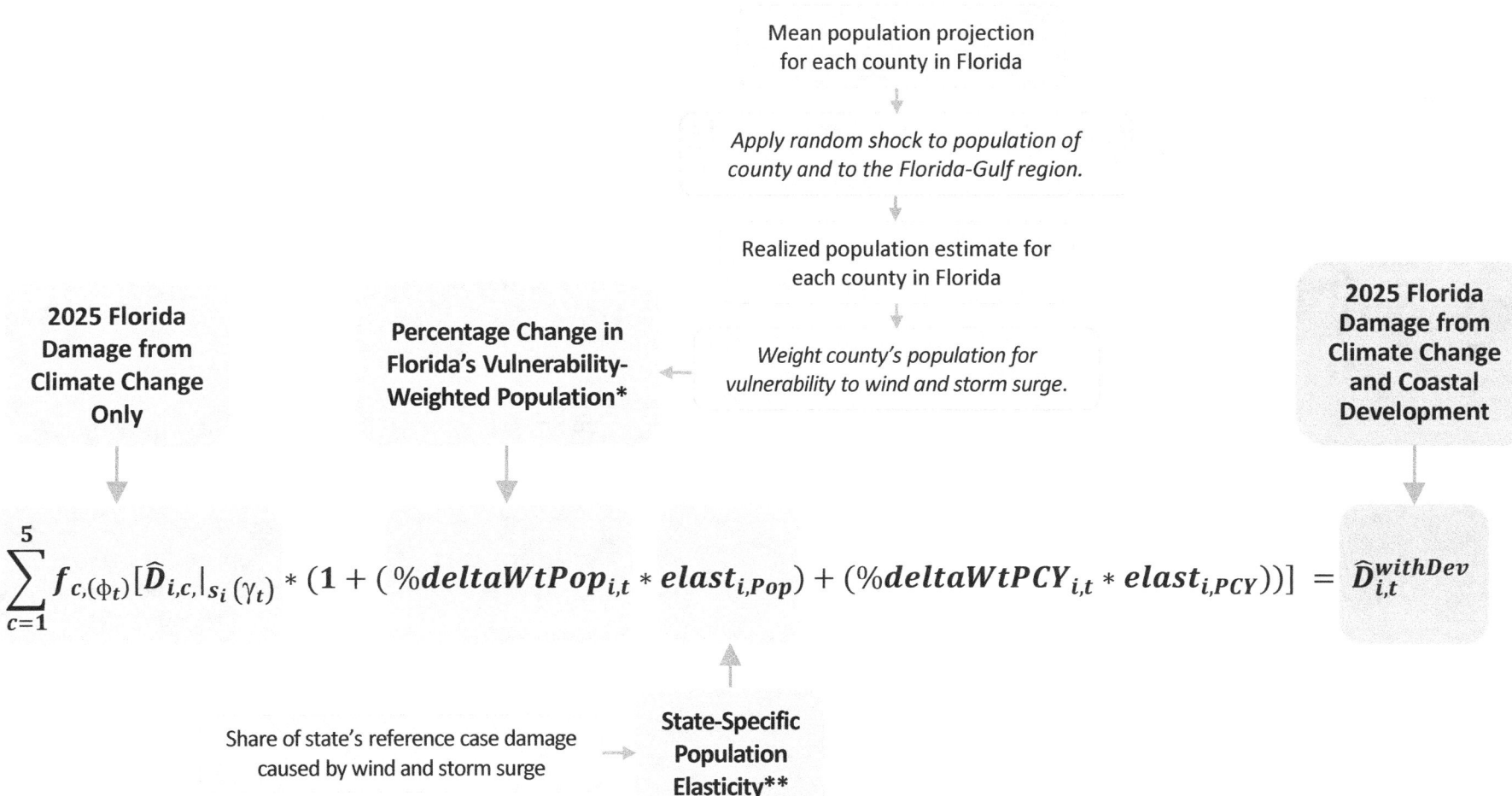

Mean population projection
for each county in Florida

*Apply random shock to population of
county and to the Florida-Gulf region.*

Realized population estimate for
each county in Florida

*Weight county's population for
vulnerability to wind and storm surge.*

**2025 Florida
Damage from
Climate Change
Only**

**Percentage Change in
Florida's Vulnerability-
Weighted Population***

**2025 Florida
Damage from
Climate Change
and Coastal
Development**

$$\sum_{c=1}^{5} f_{c,(\phi_t)} [\widehat{D}_{i,c,}|_{s_i(\gamma_t)} * (1 + (\%deltaWtPop_{i,t} * elast_{i,Pop}) + (\%deltaWtPCY_{i,t} * elast_{i,PCY}))] = \widehat{D}_{i,t}^{withDev}$$

Share of state's reference case damage
caused by wind and storm surge

**State-Specific
Population
Elasticity****

*Percentage change in Florida's vulnerability-weighted per capita income $\left(\%deltaWtPCY_{i,t}\right)$ is calculated in a similar manner.

**Population elasticity indicates the percentage change in damage given a percentage change population. CBO also estimated a state-specific PCY elasticity $(elast_{i,PCY})$.

Applying Random Shocks to Generate County Population Estimates for Each Simulation, Florida Example

County population shock $(W_{y,t})$ based on random draw from N(0,1)

Mean Population Projection for Each County in Florida

Based on projected U.S. population growth and county's share of historic U.S. population growth

Correlation Coefficient

Correlation between county and regional growth in the Florida-Gulf Region

Sea-Level-Rise-Adjusted County Draw

Adjustment slows population growth if SLR significantly increases expected damage. For example, county draw is cut in half (doubled if negative) if SLR doubles mean estimate of climate only damage in Florida.

$$\overline{Pop}_{y,t} + \sigma_{y,t}^{Pop} * \rho Pop_R * Z_{R,t} + \sigma_{y,t}^{Pop}\left(1 - \rho Pop_R{}^2\right)^{1/2} * AdjW_{y,t} = \widehat{Pop}_{y,t}$$

County-Specific Standard Error

Size-based percentage of $\overline{Pop}_{y,t}$

Florida-Gulf Region Population Shock

Based on random draw from N(0,1)

Realized Population Estimate for Each County in Florida

A similar method is used to estimate each county's per-capita income for each simulation.

Weighting County Population for Vulnerability to Wind and Storm Surge Damage

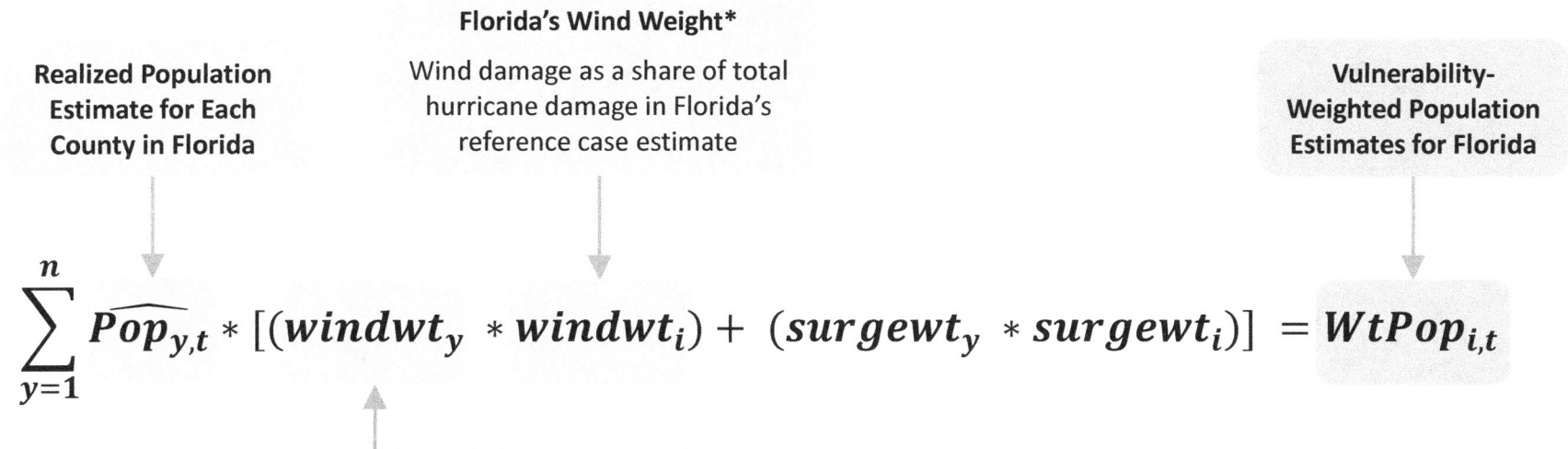

Realized Population Estimate for Each County in Florida

Florida's Wind Weight*
Wind damage as a share of total hurricane damage in Florida's reference case estimate

Vulnerability-Weighted Population Estimates for Florida

$$\sum_{y=1}^{n} \widehat{Pop_{y,t}} * [(windwt_y * windwt_i) + (surgewt_y * surgewt_i)] = WtPop_{i,t}$$

County-Specific Wind Weight*

County's share of increase in probability-weighted wind damage in Florida if $100 of additional property were added to each county

(Based on maps from the National Hurricane Center, output from FEMA's Hazus model, and RMS reference case data)

* County-specific weights for storm surge damage ($surgewt_y$) and state-specific weights for storm surge damage ($surgewt_i$) were also calculated.

A similar method is also used to estimate each county's per capita income for each simulation.

Elasticity Estimates

- Elasticity indicates a percentage change in hurricane damage for a given percentage change in population (or per capita income).

- Only a limited number of estimates are available.
  - Reflect both intentional and unintentional changes in vulnerability
  - Vary across countries

- The Bakkensen and Mendelsohn study is main source of U.S. elasticity estimates (results apply mainly to wind damage):
  - Per capita income elasticity = 1.15
  - Population elasticity not significantly different from zero

Elasticity Estimates Used in CBO's Analysis

- For wind:
  - Per-capita income elasticity = 1
  - Population elasticity = 0.25

- For storm surge:
  - Per capita income elasticity = 0.75
  - Population elasticity = 0.5

Implications of Elasticity Estimates Used in CBO's Analysis

- Doubling of both population and per capita income (roughly a 400 percent increase in GDP) would cause damage to increase by 250 percent.

- Damage due only to coastal development (holding climate constant) grows at roughly 60 percent of the growth rate of GDP.
  - Denser development can reduce:
    - Wind damage per structure (if buildings are closer together)
    - Storm surge damage per structure (if buildings are taller)
  - More expensive construction may be less vulnerable to damage.

Preliminary Damage Estimates, by Dollar Amount, Under the Scenario with Climate Change and Coastal Development

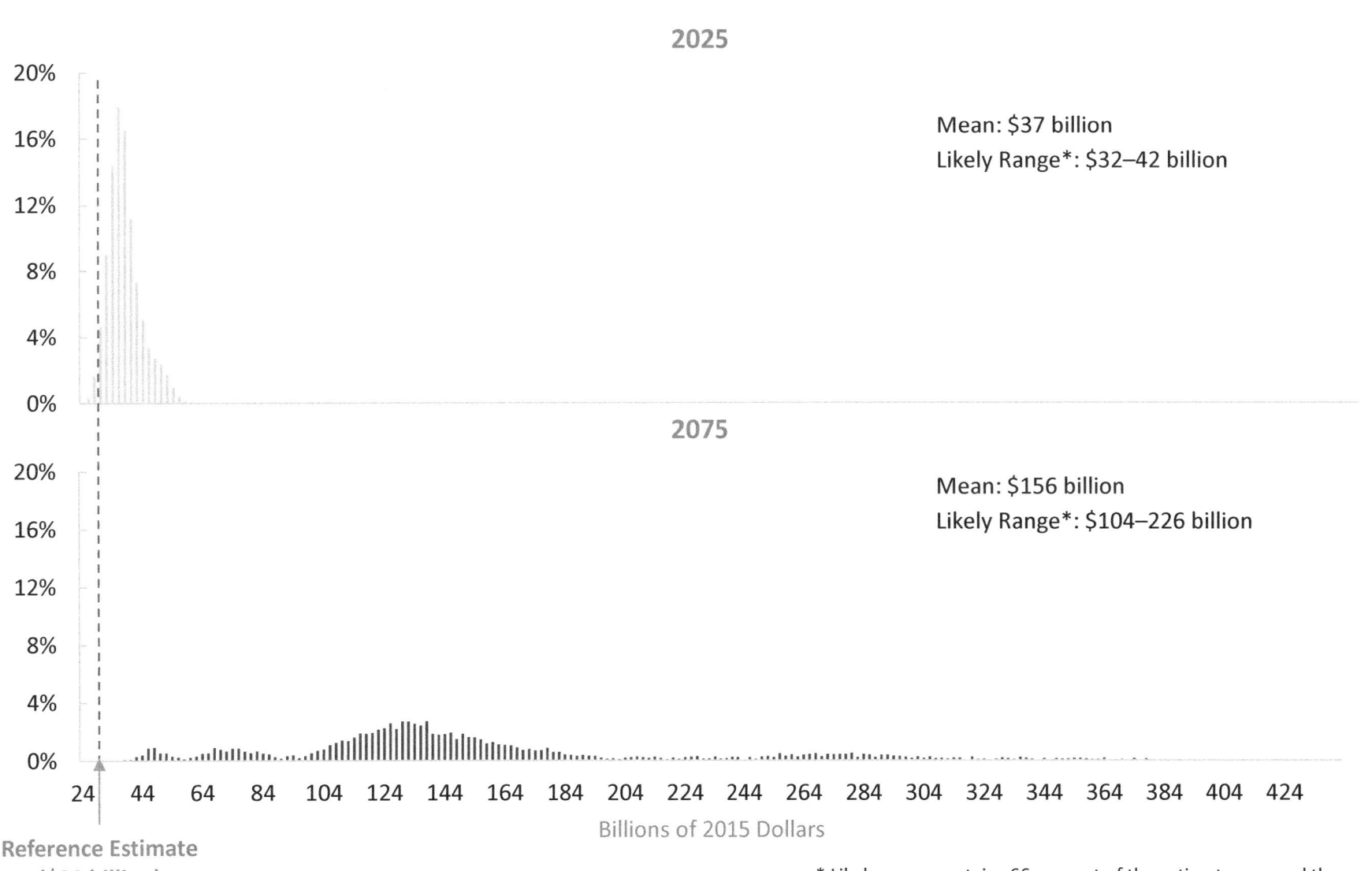

Preliminary Damage Estimates, by Share of GDP, Under the Scenario with Climate Change and Coastal Development

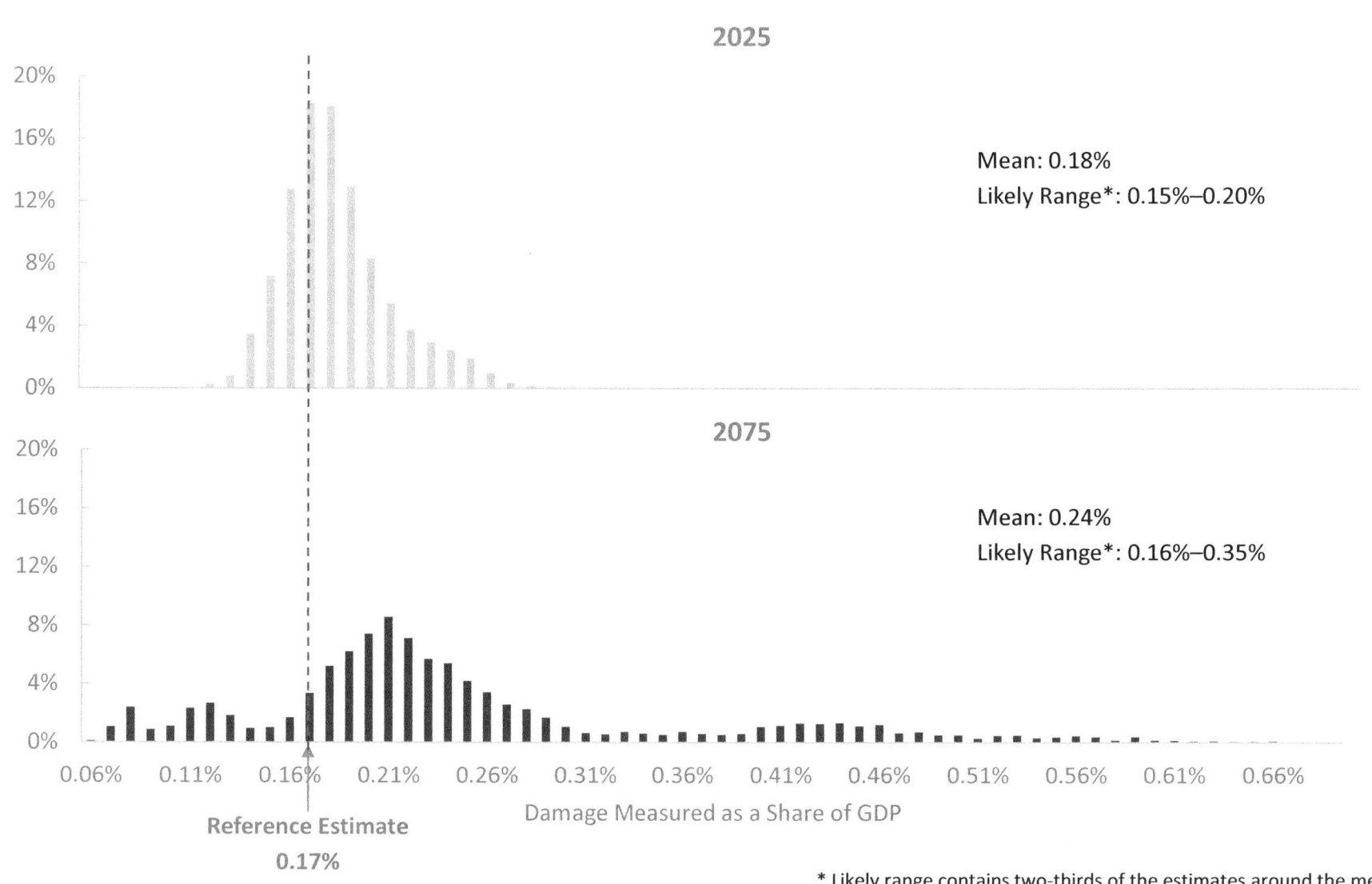

2025

Mean: 0.18%
Likely Range*: 0.15%–0.20%

2075

Mean: 0.24%
Likely Range*: 0.16%–0.35%

Reference Estimate
0.17%

Damage Measured as a Share of GDP

* Likely range contains two-thirds of the estimates around the mean.

Sensitivity Analysis Using Alternative Elasticity Estimates, Which Imply Different Levels of Adaptation

Hurricane damage estimates for 2075 under the scenario with climate change and coastal development

| | Wind Elasticities | | Surge Elasticities | | Mean Damage Estimates | | "Likely Range" of Damage Estimates | | | |
| | | | | | | | Billions of 2015 Dollars | | Percentage of GDP | |
Scenario	PCY	Pop	PCY	Pop	Billions of 2015 Dollars	Percentage of GDP	Low End*	High End**	Low End*	High End**
Base Case	1.0	0.25	0.75	0.5	156	0.24	104	226	0.16	0.35
Low Adaptation	1.25	0.5	1.0	0.75	186	0.29	124	265	0.19	0.41
High Adaptation	0.75	0	0.5	0.25	126	0.20	84	185	0.13	0.29

Notes: Reference case damage in 2015 (present conditions) = $29 billion; 0.17 percent of GDP.

Adaptation includes intentional (for example, building sea walls) and unintentional changes (for example, denser housing) that lead to reductions in damage.

Low elasticities imply a greater degree of adaptation than higher elasticities.

* Low end = 17 percentile; ** High end = 83 percentile.

PCY = per capita-income elasticity; Pop = population elasticity.

Summary

- This analysis is preliminary and results are subject to change.

- By 2075, climate change and coastal development cause expected damage to be five times greater than today (measured in 2015 dollars).
  - Likely range is three times to eight times greater

- The economy in 2075 is projected to be nearly four times larger than it is today.

- In combination, climate change and coastal development cause damage to increase more rapidly than GDP.

- In contrast, damage due only to coastal development grows more slowly than GDP.

Summary (Continued)

- The mean estimate of damage is:

  - 0.17 percent of GDP under current conditions

  - 0.24 percent of GDP in 2075

  - The increase in the mean estimate of damage as a percentage of GDP in 2075 (relative to today) accounts for less than 0.1 percent of GDP

- Estimates are uncertain.

  - Measured in 2015 dollars, the likely range in 2075 is 12 times larger than in 2025

  - Measured as a share of GDP, the likely range in 2075 is 4 times larger than in 2025

Key Sources Used in This Analysis

- Laura A. Bakkensen and Robert O. Mendelsohn, *Risk and Adaptation: Evidence From Global Tropical Cyclone Damages and Fatalities*, Working Paper (The University of Arizona, August 2014), www.ncsu.edu/cenrep/workshops/documents/Bakkensen.pdf.

- Kerry A. Emmanual, "Downscaling CMIP5 Climate Models Shows Increased Tropical Cyclone Activity Over the 21st Century," *Proceedings of the National Academy of Sciences*, vol. 110, no. 30 (July 2013), www.pnas.org/content/110/30/12219. Additional data was provided to CBO by the author.

Key Sources Used in This Analysis (Continued)

- Trevor Houser and others, *American Climate Prospectus: Economic Risks in the United States* (Rhodium Group, October 2014), Technical Appendix III: Detailed Sectoral Models, http://rhg.com/reports/climate-prospectus. Houser and others provides a description of the RMS model.

- Thomas R. Knutson and others, "Dynamical Downscaling Projections of Twenty-First-Century Atlantic Hurricane Activity: CMIP3 and CMIP5 Model-Based Scenarios," *Journal of Climate*, vol. 26, no. 17 (September 2013), http://journals.ametsoc.org/doi/abs/10.1175/JCLI-D-12-00539.1. Additional data was provided to CBO by the author.

Key Sources Used in This Analysis (Continued)

- Robert E. Kopp and others, "Probabilistic 21st and 22nd Century Sea-Level Projections at a Global Network of Tide-Gauge Sites," *Earth's Future*, vol. 2, no. 8 (August 2014), http://onlinelibrary.wiley.com/doi/10.1002/2014EF000239/full. Risk Management Solutions based its sea-level-rise projections on Kopp and others.

- Risk Management Solutions, "Catastrophe Models," www.rms.com/products/models-cat.

www.ingramcontent.com/pod-product-compliance
Lightning Source LLC
Chambersburg PA
CBHW081319180526
45170CB00007B/2777